Climate change may seem like a future problem but its effects are happening today.

TODAY'S HEADLINES

12 THINGS TO KNOW ABOUT CLIMATE CHANGE

BLACK RABBIT BOOKS

VICKI HAYES

Table of Contents

The Earth Is *Warming Up*

1

Earth's **climate** has always changed naturally. It is normal for the global temperature to go up and down. This usually happens over long periods of time. This long-term shift is called climate change. But the last 150 years have been different. Temperatures have risen faster than ever before. There are scientists called climatologists. They study the Earth's temperatures. They say human activity has added to this rapid change.

The average temperature of the world is normally 59 degrees Fahrenheit (15 degrees Celsius). The Intergovernmental Panel on Climate Change (IPCC) was created in 1988. It is part of the United Nations (UN). It looks at climate data. In the last century, the global temperature rose about 2 °F (1.1 °C). It is predicted to rise more. The IPCC believe it will rise another 3 °F (or 1.5 °C) over the next few decades. This is called global warming.

An increase of one degree can cause big changes. As the earth gets warmer, glaciers melt. Ocean levels rise. **Ecosystems** are affected. Human lives may be in danger. Communities can suffer. Coastal areas are at risk of severe flooding. Weather can be more extreme. Warmer temperatures will impact most life on Earth.

2050 Year scientists predict many species will die out.

More than 1 million plants and animals may disappear if global warming is not slowed. • Global warming can make droughts more frequent. • It can make floods more severe. • Higher temperatures are bad for fish.

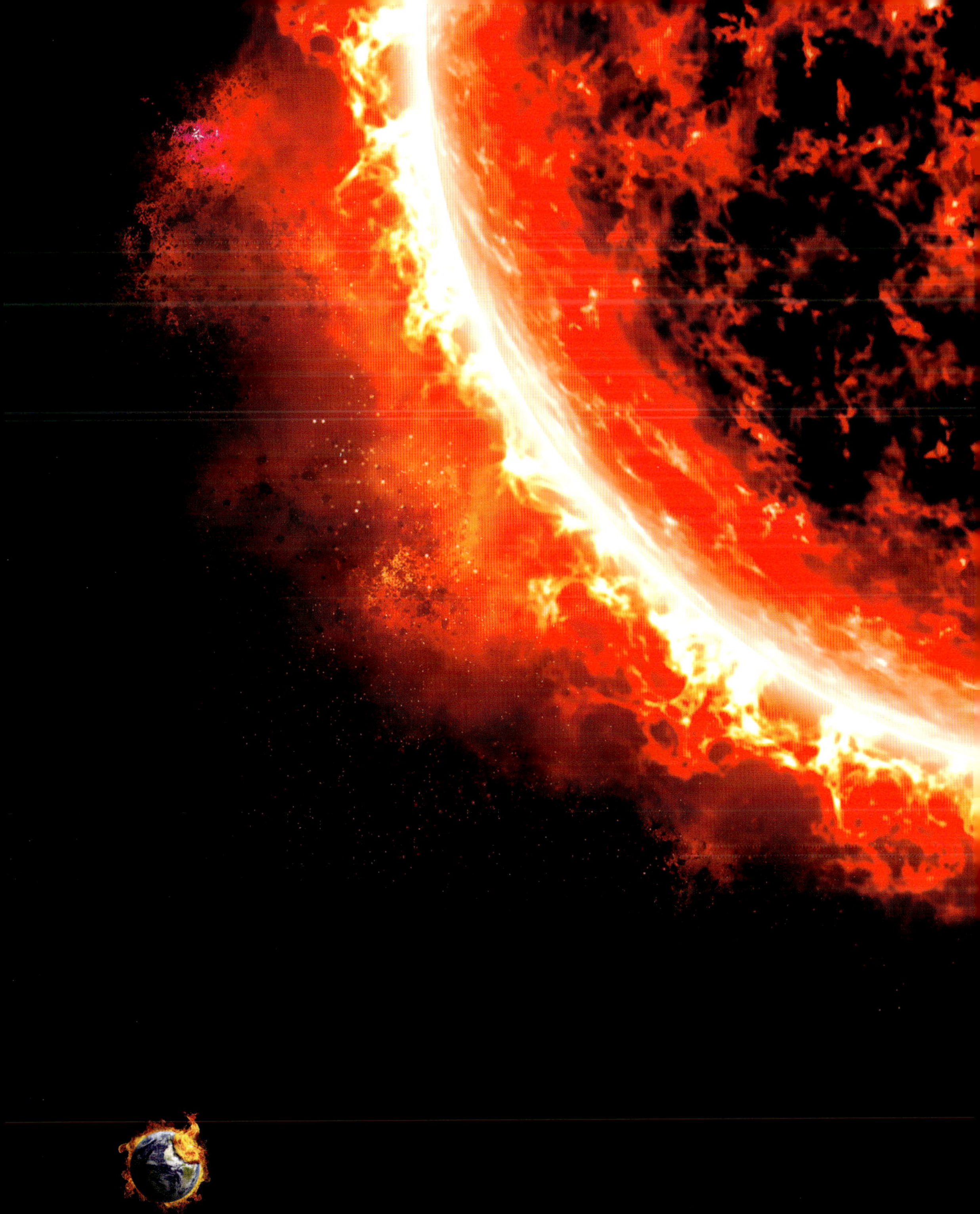

With global warming, Earth traps more of the Sun's heat.

Climate Change Is *Real*

2

Scientists agree. We must slow global warming. Most scientists believe humans are the main cause of climate change. However, there are some people who disagree. They think climate change is not serious. They are called skeptics.

The skeptics are not sure humans are to blame. They don't think scientists know enough. Global weather patterns are complicated. Earth has been through climate changes before. There have been ice ages and warming periods. Natural disasters, such as volcanic eruptions, have also changed Earth's climate.

However, there are a lot of facts about climate change. Studies began in the 1970s. They tracked the global temperatures. They show that Earth is warming. It is happening faster than at any time in the last

Deserts around the world are growing hotter and dryer.

10,000 years. The IPCC agrees. It says the influence of humans on global warming is now a fact.

50 Number of climate change indicators used by the Environmental Protection Agency.

Indicators include rising sea levels, melting glaciers, and warming oceans. • Scientists use satellites and new technologies to gather information. • They have collected data for many years.

Think About It

Have you ever had to prove something? Did you have to find data to support your statement? Think of something you want to prove and plan how you would do it.

Greenhouse Gases Trap *Heat*

3

Earth's **atmosphere** is made of gases. Some are called **greenhouse gases**. We need these gases to survive. They trap heat from the sun. Without them, the planet would be an average -4 °F (-20 °C). That's too cold to grow food. We wouldn't be able to raise animals. These gases are helpful for keeping our world livable.

However, too much of these gases is also a problem. They trap too much heat. Earth's surface cannot cool down enough at night. Then the temperature of the air and oceans rise. This is what is happening today.

Greenhouse gases have natural sources. Humans and animals breathe out carbon dioxide. Rotting plant matter gives off methane gas. This is a natural process. But today, greenhouse gases are increasing too fast.

Exhaust fumes from cars is a leading cause of climate change.

They are causing big changes in climate. Many living things will not be able to adapt fast enough.

Carbon dioxide (CO_2) is the most common greenhouse gas. CO_2 levels have gone up 50 percent in the past 200 years. Methane is another gas. It is not as common. But it traps more heat than CO_2. Methane levels are two and a half times what they were in the 1600s. These gases come from several sources. Energy production is one. Transportation is another. Transportation is the largest source in the United States.

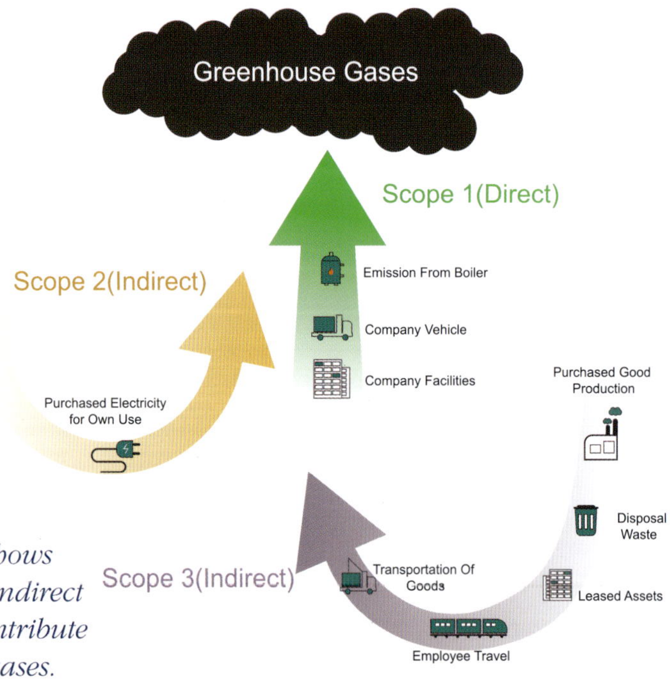

This diagram shows the direct and indirect ways people contribute to greenhouse gases.

28 Percentage of US greenhouse gas emissions from transportation.

The second largest US source is power plants. • The US and China are the world's largest producers of greenhouse gases. • Methane is responsible for 25 percent of today's global warming.

Humans and animals exhale CO_2 but not enough to impact the climate.

Fossil Fuels Release *Harmful Gases*

4

Fossil fuels are energy sources. They are found in the earth. Oil, natural gas, and coal are fossil fuels. They were formed millions of years ago. Long ago, dead plants and animals sank in mud. They were buried under layers of rock. Slowly, the pressure of the rocks changed them. They turned into fossil fuels.

Fossil fuels are a limited **resource**. New supplies can be found. But there aren't many. Fossil fuels are burned for energy. This process releases CO_2. CO_2 is the biggest contributor to global warming. Nearly 90 percent of CO_2 comes from fossil fuels.

During the last 150 years, fossil fuel **emissions** have increased. Today, about 80 percent of power in the United States comes from these fuels. This energy heats and cools homes and businesses. It powers electronics.

Machines drill for oil then transport it along a pipeline or in a tanker truck.

200 Years that CO_2 can last in the atmosphere.

China produces the most CO_2 each year. • The United States produces the second most. • Over 60 percent of electricity is produced by fossil fuels.

Oil is sold by the barrel.

It runs cars and factories. Fossil fuels are a large part of our lives. It will take time to switch to other energy sources. To help, we need to **conserve** energy. This will reduce emissions.

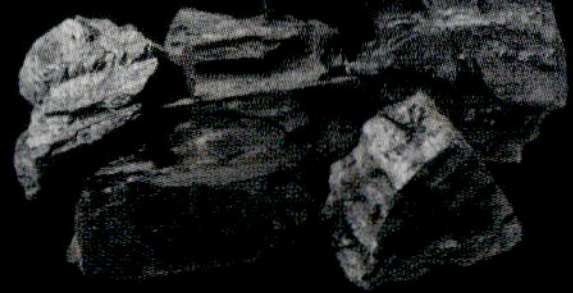

A digger scoops up coal, which is burned for fuel. Coal burning contributes to climate change.

CARBON SINKS The earth removes CO_2 from the air naturally. It uses something scientists call a carbon sink. Oceans and trees are carbon sinks. CO_2 dissolves in cold oceans. Then, it sinks to the bottom. Trees absorb the gas and use it to grow. Without carbon sinks, the effects of climate change would be much worse.

Farming Affects *the Climate*

5

Before fossil fuels, farms were small. They averaged around 100 acres (40 hectares). Farmers worked by hand to raise livestock and grow crops.

Then about 150 years ago, things changed. Fossil fuels started powering farm machinery. More work could be done with machines than by hand. Farms became much bigger. Today's farms average 446 acres (181 ha). Over half the land in the United States is used for farming.

The world's population has quadrupled in the last century. Farming practices have changed. Now more food can be produced. But many of the new practices affect climate change. Fertilizers give off a lot of greenhouse gas. Livestock create a lot of methane gas. Farm machinery produce carbon dioxide.

The Industrial Revolution had a big impact on scale of farming.

Global food systems create one-third of greenhouse gases. Wasted food makes up half of that. Food rots in landfills. It releases methane.

A row of 22 harvesters work to harvest a crop of soybeans in Brazil.

44 Percentage of land used for food production.

Two thirds of that is used for grazing livestock. • Over 10 percent of US greenhouse gases come from farms. • About one-third of food grown worldwide is wasted.

Livestock's affect on the environment is a big reason some people eat vegetarian diets.

COW POWER

A single cow can produce 65 pounds (29.5 kg) of manure a day. As the manure decomposes, it releases methane. This adds to greenhouse gases. But a new technology turns that methane into electricity. Then, it can be used for lights in homes. It can fuel a water heater or stove. It can power farm equipment and furnaces.

Deforestation Changes *the Climate*

6

Forests are being cut down at rapid rates worldwide. This is known as deforestation. Between 2015 and 2020, about 25 million acres (10 million ha) of trees were cut down each year. The Earth needs trees. They help slow climate change. All plants absorb CO_2. But trees do it best. One tree can absorb 1 ton (0.9 metric tons) of CO_2 in its lifetime. Trees help keep Earth's temperature steady.

Wildfires destroy forests. There have been more wildfires in the last 20 years. They have been bigger than in the past, too. In 2022, over 16 million acres (6.5 million ha) of trees burned. Humans are also cutting down forests. They want to make space for crops and livestock grazing. They use a method called slash-and-burn. This clears large areas quickly. The ash from the fires fertilizes

the land. But this lasts for only a few years. Then the process has to be repeated elsewhere.

Some wildfires are natural, but human-caused ones can be very destructive.

Deforestation is bad for the Earth. Trees hold soil in place. They protect groundwater. Trees provide food and shelter for other plants and animals. Without trees, the sun dries up the soil. Rainfall decreases. In time, forests become dry like deserts.

Trees work to make a healthy planet. They put **water vapor** back into the air. They filter drinking water. Without enough trees, there will be less rain and higher temperatures. There will be more severe weather. And there will be more droughts and heat waves.

1.6 billion Number of people who use forests for shelter, water, and food.

Over 3 million species are threatened by deforestation in the Amazon. • About 95 percent of deforestation occurs in the tropics. • Brazil has the highest rates of deforestation.

Carbon dioxide and oxygen are important gases to plants and animals.

Sea Levels Are *Rising*

7

Scientists studied sea levels in the 20th century. They found that sea levels had risen 8 inches (20 cm) since 1900. If greenhouse gases increase, sea levels will rise more. They could rise over 7 feet (2.1 m) by the year 2100. Coastal cities would flood. Around 70 million people would lose their homes.

As the temperature rises, the oceans heat up. They absorb about 80 percent of the extra heat. Warmer temperatures also melt glaciers. Most glaciers are found in cold, high places. They are in mountains and polar areas. Glaciers move slowly. They slide down slopes. The melted runoff flows into oceans. This causes sea levels to rise.

Melting ice sheets cause the same problem. The largest is the Antarctic ice sheet. The second largest is in

15 Percentage of greenhouse gases absorbed each year by the Southern Ocean.

CO_2 makes the oceans more **acidic**. This destroys the shells of sea life. • Latin America and the Caribbean are at the highest risk of flooding.

An exaggerated image illustrates the serious effects of sea levels rising.

Greenland. It covers 80 percent of the country's surface. If this ice sheet completely melted, sea levels could rise by 23 feet (7 m). The effect of this ocean rise would be felt in the southern hemisphere. This is because Greenland's gravity would change. Without ice, it would weigh less. The melted ice water would slosh to parts of the world where gravity is stronger.

Polar bears are just one animal threatened by climate change.

POLAR BEARS IN PERIL

Climate change threatens polar bears. Arctic ice is shrinking. This affects the bears' hunting for food. They use sea ice to catch seals, their main prey. There are 26,000 bears in the Arctic. Scientists think one-third of them will be gone by 2050.

8 Countries Are *Adapting*

The effects of climate change are being felt everywhere. Sea levels are rising at five times their previous rate. Millions of people are threatened by coastal flooding. People must adapt. Otherwise, they are at risk. Some countries are adapting to climate change. They are using new technologies.

Venice, Italy, has a system of flood barriers. They built them in 2020. There are 21 flood gates. They can protect the city from tides up to 10 feet (3 m) high. In 2017, Bangkok, Thailand, built an 11-acre (4.5-ha) park. The park will capture flood water. It will be funneled into a pond. Under the park are storage tanks. Altogether, the park and tanks can hold 1 million gallons (3.8 million liters) of water. The Maldives are in the Indian Ocean. There are more than 1,000 islands. They are building a floating city. It will be a series of units. They will rise and fall with the sea level. It is planned for 2028.

Small communities are also dealing with flooding. They raise roads and move low-lying buildings. They stabilize shorelines. They restore wetlands and other ecosystems. Wetlands trap excess water. They slow the spread of flood waters.

$14.2 trillion Amount of money needed to recover from coastal flooding by the end of 2030.

Hundreds of big cities will face increased flood risk over the next 25 years. • California requires coastal communities to have a plan for sea level rise. • Mangrove trees and coral reefs both help to limit flooding.

The city of Venice tests their flood barriers.

DELTA WORKS The Netherlands suffers from frequent flooding. The Delta Works was built between 1950 and 1997. It was built to protect the Netherlands for 100 years. It has three locks, six dams, and four storm barriers. When the gates are closed, the dam is watertight. But sea levels are still rising. A fifth barrier was added in 2015.

A storm barrier is raised in the Delta Works.

The Arctic Is Warming *Faster*

9

The northern Arctic has a harsh, cold climate. It has treeless plains called tundra. Beneath the tundra is permafrost. This is permanently frozen rock or soil.

Scientists know the Arctic is melting. Satellite data shows the sea ice has shrunk. It has shrunk every month since 1979. When permafrost thaws, it releases methane. Much of Russia is in the Arctic. About 65 percent of Russia is covered with frozen tundra. This area is warming 2.5 times faster than the rest of the world. Large amounts of gases are released from the thawed soil there.

The Arctic region contains the Arctic Ocean. This ocean touches eight countries. Humans have lived there for thousands of years. They have adapted their way of life to the climate. Warmer temperatures have forced people to change their lives. They use boats instead of dog sleds or snowmobiles.

Narwhal whales only live in the Arctic Ocean.

A warming Arctic has affected the caribou's diet and habitat.

4 million Number of people who live in the Arctic region.

Ice reflects the sun. Less ice means less heat is reflected back into space. • This means more heat warms up the Earth's surface. This leads to more heat waves. • Arctic wildfires will be more frequent and more extreme.

Many Arctic people rely on whales, seals, caribou, and fish for food. Climate change threatens their food supply. It threatens their jobs. It also threatens Arctic animals. Polar bears are **endangered**. So are some caribou and narwhals. The most endangered animal in the region is the Arctic fox.

Climate change can affect how many fish fisherman can catch.

THINK ABOUT IT

What are some ways native people in the Arctic are affected by climate change? Why do you think they might not want warmer weather?

Climate Change Affects the *Weather*

10

Climate change alters weather patterns. Some of these changes are extreme. Extreme weather includes heat waves and droughts. It includes floods, blizzards, and wildfires. These are all happening more often. And scientists say it will get worse.

Warm air creates more storms. It affects **precipitation**. There is more water vapor in the air. Rainfalls are heavier. In 2021, extreme flooding hit Europe. Rivers flooded. Levels were higher than they had been in hundreds of years. The rain fell too fast. Even small streams became threats. Bridges collapsed. Cities lost power and cell service. Hundreds of people were missing.

Hurricanes are strong tropical storms. They have heavy rain and strong winds. They can damage buildings, trees, and cars. Climate change is making hurricanes wetter and windier. They are more frequent. In 2023, there were 28 weather disasters in the United States that cost over $1 billion each. They included 17 severe storms, 2 tornado outbreaks, and 2 tropical cyclones. There were floods along the East Coast, a winter storm in the northeast, and a wildfire in Hawaii.

Heat waves are becoming more frequent. US cities once averaged two heat waves a year. In the 2020s, the US had more than six heat waves a year.

30 Number of named hurricanes in the Atlantic in 2020. That is a record.

The increase in tropical storms is related to the increase in sea temperatures. • Heat records were broken all over the world in 2023. • Nine of the ten warmest years have occurred in the last 25 years.

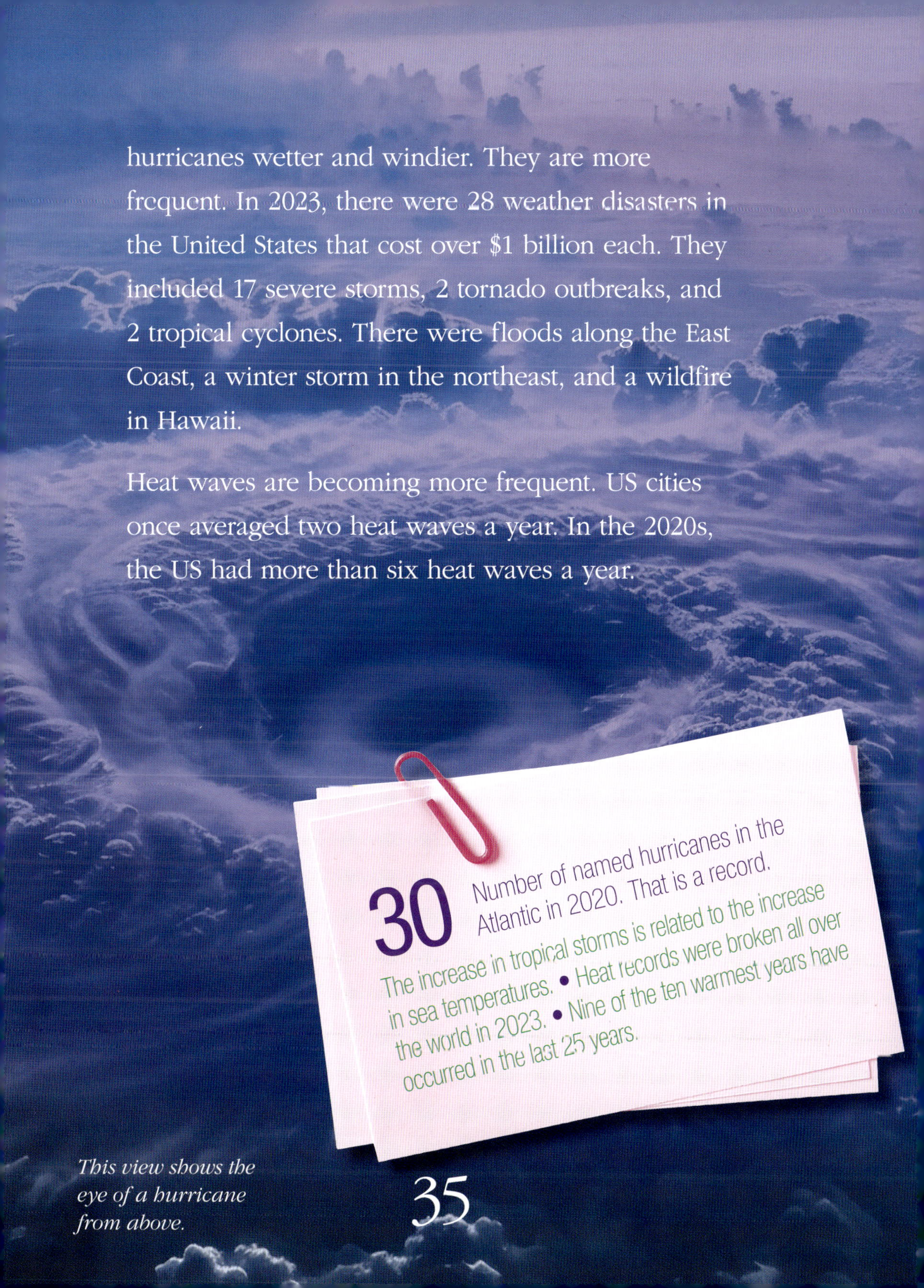

This view shows the eye of a hurricane from above.

CLIMATE VS. WEATHER

Is it cloudy? Sunny? Rainy? These are all weather questions. Weather is the way the atmosphere behaves each day. Climate is the long-term pattern of weather in a certain area. For example, the climate of Arizona is warm with low humidity. Weather can be predicted. But it changes every day. Climate stays the same day to day.

Extreme weather can create destructive tornadoes.

Climate Change Alters *Ecosystems*

11

Climate change harms whole ecosystems. Weather patterns change. Habitats are warmer. Many plants and animals struggle to adapt. Some animals have moved farther north to cooler areas. Some have moved into deeper waters. Many animals have changed their behaviors. In some places, spring arrives sooner. This causes birds to breed and migrate earlier.

Life that cannot adapt will die out. Many plants and flowers rely on insects to spread their pollen. Climate change can cause the plants to bloom early. Insects might not be active then. Without insects, the plants cannot multiply. They may decrease in large numbers. They may even die out completely.

Animals and insects that depend on these plants will suffer. For example, koalas eat mainly eucalyptus leaves.

These trees require cool climates to grow. But heat waves and droughts are threatening the trees. The number of trees may decline. Then more koalas will starve. Scientists must find ways to reduce greenhouse gases. If not, thousands of species may die out.

A koala munches on eucalyptus leaves.

Monarch butterflies migrate south to Mexico each winter.

60 Percentage of insects trying to get in sync with the plants they rely on.

Climate change is increasing the threat of **invasive species**. • Climate change affects food webs. • Changes in seasons cause migratory animals to travel sooner. Food and habitats are harder to find.

Humans Can Make a *Difference*

12

The global climate is a complex topic. Data about weather patterns changes constantly. There is much information to study. Most scientists agree that the earth is warming. What can humans do about it?

There are ways to reduce its effects. Even small changes can make a difference. Reduce, reuse, recycle is a useful phrase. It reminds people to reduce their waste. People can also reuse and recycle items to avoid wasting limited resources.

Driving a car produces greenhouse gases. Using electricity from fossil fuels also creates these gases. Even throwing away garbage produces them. People can help by using public transportation. They can use **renewable energy**. They can recycle trash. They can **compost** garbage.

Solar panels collect light from the sun and turn it into energy.

Recycling, reusing,
and reducing waste
can help protect our
environment.
REDUCE
REUSE
RECYCLE

Governments around the world are looking for ways to help. They want to limit the damage of climate change. The United States is already using some renewable energy sources. This energy will someday replace fossil fuels. Buildings can save energy by insulating their walls. Solar panels can turn sunlight into energy. Saving forests can help. Planting new trees would allow more CO_2 to be absorbed. These actions can slow down climate change.

2023 Year the UN Climate Change Conference met in Dubai.

Members laid the groundwork to end use of fossil fuels. • The UN shared 10 ways people can help slow climate change. • Some ways include saving energy at home and throwing away less food. • The UN also suggests eating more vegetables and planting native species.

Think About It How can you help to slow climate change? Are there ideas on these pages that you can do? Find out what your community or school is doing and join in.

How Can

Reduce Fossil Fuel Use

Humans rely on fossil fuels for transportation. This creates greenhouse gases. To help, carpool or use public transportation. These are more efficient ways to travel. When you can, walk or ride a bike.

Plant Trees and Native Plants

Trees are important absorbers of CO_2. And native plants are better for the environment. They are adapted to the area's climate, so they are hardier. They usually don't need as much water to grow.

Make Compost

Don't throw your food scraps away. Instead, create a compost. Look for instructions online. You can start with a small bucket in your kitchen. Many communities have big bins you can throw your scraps into. Compost is great for gardens.

You Help?

Use Nontoxic Chemicals

Cleaning chemicals are often toxic for the air and water. These include laundry detergents, cleaning products, air fresheners, shampoos, and even candles. Choose eco-friendly products. Your home will be clean and healthier.

Reuse and Upcycle

Swap clothes and games with friends. Buy something used instead of brand new. And repurpose old things to make them new again. This is called upcycling.

Spread the Word

Talk to others about the changes you are making. Explain how they are helpful to our planet. You can encourage family members and friends to make a change as well.

Glossary

acidic
Containing a chemical substance that can dissolve things.

atmosphere
The gases that surround a planet.

climate
The usual weather conditions in a particular place or region.

compost
Decaying organic material, such as leaves and grass, used to improve soil.

conserve
To use something carefully in order to prevent loss or waste.

ecosystem
A community of living things in one place.

emission
Something that is released into the air.

endangered
Close to no longer existing.

fossil fuel
A fuel that is formed in the earth from dead plants or animals.

greenhouse gas
A gas in Earth's atmosphere that traps heat.

invasive species
Animals or plants that spread through an area where they are not normally from, often causing problems for native species.

precipitation
Water that falls to the ground, such as rain or snow.

renewable energy
Energy from a natural source that is refreshed at a higher rate than it is used.

resource
A supply of something that can be used when needed.

water vapor
Water in its gaseous state.

For More Information

Books

Dorey, Martin. *Kids Fight Climate Change.* Somerville, MA: Candlewick Press, 2022.

Mittermeier, Cristina. *Rebel Girls Climate Warriors: 25 Tales of Women Who Protect the Earth.* Mankato, MN: Black Rabbit Books, 2025.

Woodward, John. *Eyewitness Climate Change.* New York: DK Publishing, 2021.

Websites

NASA: Climate Kids
www.climatekids.nasa.gov

US EPA: A Student's Guide to Global Climate Change
archive.epa.gov/climatechange/kids/index.html

About the Author

Vicki Hayes has a master's degree in early childhood education and more than 30 years of experience teaching children of all ages. She currently works as a teacher and writer.

Index

• Top Rank is an imprint of Black Rabbit Books. • Edited by Alissa Thielges | Designed by Danny Nanos • Photographs © Dreamstime/Planetfelicity, 30, Dezzor, 43, Gillian Hardy, 39, Justlight, 35, Lakhesis, cover, 1; Shutterstock/Ad Gr, 7, Andrey_Popov, 44, beast01, 45, Belinda Pretorius, 48, Byjeng, 16, Christian Roberts-Olsen, 22, creativevikas, 16, Don Landwehrle, 6, FloridaStock, 26, Francesco Scatena, 20, 23, freie kreation, 4–5, GaudiLab, 32–33, grapestock, 44, Hamara, 2–3, 15, Jolygon, 46–47, Katvic, 10, Kelvin H. Haboski, 18–19, klenger, 2, Lamyai, 9, LeStudio, 36, Ljupco Smokovski, 42, LumenSt, 45, Martin Capek, 7, Master_shifu, 12, Maximillian cabinet, 8–9, New Africa, 44, Nuttapong, 17, Ody_Stocker, 11, petrmalinak, 41, photo.ua, 27, Pictures of your Life, 40, Pinkyone, 17, pixelschoen, 25, RanjanaK, 41, 45, Reynaldo Graca Lopes, 38, Roman Mikhailiuk, 34, Ruud Morijn Photographer, 29, Suranyi Mate, 13, Tatiana Popova, 14, Tools Konten, 37, TrotzOlga, 11, 21, tuntekron petsajun, 7, Vaclav Volrab, 22–23, Veroja, 31, Zaltrona, 28, TSViPhoto, 15 • Printed in China

Library of Congress Cataloging-in-Publication Data Names: Hayes, Vicki C., author. | Title: 12 things to know about climate change / by Vicki Hayes. | Description: Mankato, MN: Black Rabbit Books, [2025] | Series: Today's Headlines | Includes bibliographical references and index. | Ages 9–13 | Grades 4–6 | Identifiers: LCCN 2024020329 | ISBN 9781645823841 (library binding) | ISBN 9781645824060 (paperback) | ISBN 9781645824282 (ebook) | Subjects: LCSH: Climatic change—Juvenile literature. | Classification: LCC QC903.15 .H39 2025 | DDC 551.6—dc23/eng/20240510 | LC record available at https://lccn.loc.gov/2024020329